Débora Rafaelly Soares Silva
Taciano Pessoa
Fabiana Pimentel Macêdo Farias

USING SESAME WASTE

Débora Rafaelly Soares Silva
Taciano Pessoa
Fabiana Pimentel Macêdo Farias

USING SESAME WASTE

PRODUCTION OF FLOUR AND FOOD PRODUCTS

ScienciaScripts

Imprint

Any brand names and product names mentioned in this book are subject to trademark, brand or patent protection and are trademarks or registered trademarks of their respective holders. The use of brand names, product names, common names, trade names, product descriptions etc. even without a particular marking in this work is in no way to be construed to mean that such names may be regarded as unrestricted in respect of trademark and brand protection legislation and could thus be used by anyone.

Cover image: www.ingimage.com

This book is a translation from the original published under ISBN 978-620-6-76061-0.

Publisher:
Sciencia Scripts
is a trademark of
Dodo Books Indian Ocean Ltd. and OmniScriptum S.R.L publishing group

120 High Road, East Finchley, London, N2 9ED, United Kingdom
Str. Armeneasca 28/1, office 1, Chisinau MD-2012, Republic of Moldova, Europe
Printed at: see last page
ISBN: 978-620-7-74366-7

Copyright © Débora Rafaelly Soares Silva, Taciano Pessoa, Fabiana Pimentel Macêdo Farias
Copyright © 2024 Dodo Books Indian Ocean Ltd. and OmniScriptum S.R.L publishing group

UTILISING SESAME WASTE: PRODUCING FLOUR AND FOOD PRODUCTS

Débora Rafaelly Soares Silva

PhD in Process Engineering from the Federal University of Campina Grande. Substitute Professor at the Federal University of Campina Grande - Sumé Campus. E-mail: profdeborarafaelly@gmail.com

Taciano Pessoa

PhD in Process Engineering from the Federal University of Campina Grande. Professor at the Federal Institute of Education, Science and Technology of Rio Grande do Norte. E-mail: taciano.pessoa@gmail.com

Fabiana Pimentel Macêdo Farias

PhD in Process Engineering from the Federal University of Campina Grande. Professor at the Federal University of Campina Grande - Sumé Campus. E-mail: fabiana.pimentel@professor.ufcg.edu.br

SUMMARY

SUMMARY

The cake obtained from the sesame oil extraction process, due to its high protein content, can be used as an alternative source for developing and enriching food products. The aim of this study was to utilise the waste from sesame oil extraction to obtain flour and to study the influence of flour concentration on the chemical, physical, physico-chemical and sensory characteristics of the food produced. Sesame residue flour (SRF) was obtained from the residue (cake) generated during the sesame oil extraction process. Four formulations were used to prepare the bread and biscuits: the standard formulation (control), produced with wheat flour and 0% sesame residue flour (SRF), and the formulations with wheat flour and SRF incorporated in proportions of 5, 10 and 15%. Chemical and physico-chemical analyses were carried out on the sesame residue flour (SRF) and the processed flour products (bread and biscuits). Texture and sensory acceptance analyses were also carried out on the bread and biscuit samples. The results indicated that the breads and biscuits of the standard formulation and those added with 5% GFR obtained a higher acceptance rate than the other samples evaluated. The increase in the partial replacement of wheat flour with sesame residue flour in the bread and biscuit formulations interfered with acceptance among the judges, making the product less attractive to consume.

Keywords: Sesamum indicum L., characterisation, sensory, bread, biscuit

1. INTRODUCTION

The bakery industry has sought to meet the growing consumer demand for healthier products, especially in relation to functional foods, due to the beneficial effects they promote to health, and one of the strategies adopted is the partial or total replacement of wheat flour with wholemeal flours extracted from various plant species (SILVA et al., 2020). Due to its nutritional potential, sesame residue flour is an excellent alternative for partially replacing wheat flour in the formulation of food products, enabling the nutritional enrichment of foods such as breads, cakes, biscuits, among others. As stated by Silva et al. (2020), supplementing bakery products, such as bread and biscuits, with sesame residue flour can generate a number of benefits, as well as producing a food that is practical to consume with a high added nutritional value, and also reducing production costs for the agro-industry, adding value to this residue that would otherwise be discarded.

However, it is necessary to determine the maximum amount that can be incorporated by these residual flours when replacing traditional raw materials in food production, so that it does not negatively interfere with the product's acceptability by the consumer market.

Given the need for information on the possible changes caused by the incorporation of sesame residue flour into the quality attributes of breads and biscuits, the aim of this study was to utilise sesame oil extraction residues to obtain flour and to study the influence of flour concentration on the chemical, physical and physical-chemical characteristics, as well as the instrumental texture profile and acceptability of breads and biscuits enriched with different concentrations of FRG.

2. REVIEW BIBLIOGRAPHY

2.1 Pie

The process of extracting vegetable oil, generally obtained by pressing, solvent extraction and subsequent purification and refining, leaves a defatted residue called defatted flour, also known as bran or cake. The origin of this waste is the cleaning, preparation and milling of grains, berries and kernels (MMA, 2006).

The sesame residue or cake obtained from the oil extraction process can be used for human and animal food without any restrictions, and has a high protein content of 40 to 50 %, depending on the process used. If obtained using the Expeller method (pressing the grains), the cake has an average water content of 8.2%, 12.8% oil, 22.8% carbohydrates, 11.8% ash and a low fibre content of 4.7%. (MILANI; GONDIM; COUTINHO, 2005); with high levels of B vitamins and a high concentration of sulphur-containing amino acids, especially methionine (1.48%), which is two to three times higher than that found in soya, cotton and peanut cake (EMBRAPA ALGODÃO, 2006).

This waste is also rich in fibre and minerals, of which calcium stands out, with around 1500mg in 100g of semi-fatted cake (ARRIEL; VIEIRA; FIRMINO, 2006). Semi-degreased pie can be considered a source of fibre, because in Brazil, Normative Resolution no. 27 of the National Health Surveillance Secretariat establishes, in the technical regulation on complementary nutritional information, that a food can be considered a source of dietary fibre when the finished product contains 3g/100g of fibre for solid foods (BRASIL, 1998). Compared to the FAO standard, defatted sesame protein has an adequate amino acid composition, except for a slight deficiency in lysine and methionine (FIRMINO, 1996).According to MAIA et al. (1999), although sesame waste is not a conventional food, it can contribute to meeting the protein and energy needs of human groups. In addition, milling the defatted sesame cake produces a flour with an excellent texture and a very light colour.

2.2 Sesame residue flour (FRG)

Defatted sesame flour (DFS), a by-product of oil extraction, contains an average of 50 per cent protein because the seed is rich in sulphur amino acids, a rare characteristic among proteins of plant origin (QUEIROGA & SILVA, 2008). When sesame flour is intended for human consumption, special attention must be paid to the quality of the raw material and its processing, in order to ensure the safety, health, quality and nutritional value of the extracted cake (SUBRAMANIAN, 1980).

Defatted sesame flour has a high protein value and therefore has the potential to be used in various formulations (MAIA et al., 1999). It can be used in the preparation of breads, cakes and biscuits, enriching foods in order to meet the population's nutritional needs.

Maia et al. (1999) worked with a defatted protein mixture obtained from sesame, with extruded flour from caupi (Vigna unguiculata L. Walp) in order to check the nutritional value of the mixture. The formulated food was considered to be of good nutritional quality.

Figueiredo & Modesto Filho (2008) suggest in their studies that the intake of sesame flour in the diet can make a beneficial contribution to reducing the risk of diabetes and obesity, as well as helping to control the glycaemic profile and weight in type 2 diabetic patients.

Nascimento (2010), working on the production of corn products enriched with grains and semi-fatted sesame cake by thermoplastic extrusion, found that this incorporation can bring great health benefits. Finco et al. (2011) observed that the addition of sesame flour to yoghurt promoted a sensorially acceptable product, rich in proteins and lipids.

2.3 Bread

According to Resolution RDC no. 263 of 22 September 2005, breads are products made from wheat flour and/or other flours, and may contain other ingredients, as long as they do not detract from the product. They can have a variety of toppings, fillings, shapes and textures, and can include expressions relating to the ingredient that characterises the product (BRASIL, 2005). As recipes improved in search of new flavours, other ingredients were added to the recipe, such as spices and seeds or something else that characterised the bread's distinctive flavour.

Replacing wheat flour with other grains can cause changes in the structural behaviour of the dough. The changes occur in the characteristics of the dough, the fermentation time and the final quality of the product. However, the percentage of substitution is directly related to the changes that can occur, such as the dilution of the gluten-forming protein present in the wheat flour.Thus, adjusting the mixing time of the dough, fermentation time and temperature, the manufacturing process and the quality of the wheat flour can minimise some of this effect (EL-DASH & GERMANI, 1994). Many studies have been carried out on mixed flours with a view to their application in the bakery and pastry sector. Wheat grains (SILVA, 2007), quinoa, soya, oats, wheat bran (SHENOY & PRAKASH, 2007), oat bran, linseed (KOCA & ANIL, 2007), sesame and sunflower seeds, rye, as well as leaves (BORNEO & AGUIRRE, 2008), fruit and other parts of these plants are being studied.

Bread is one of the most widespread foods in our country and has become one of the main calorific sources in the Brazilian diet. However, wheat proteins are of poor nutritional quality due to the deficiency of the amino acid lysine (one of the components of the protein), which limits the utilisation of its protein by the human body by approximately half. On the other hand, sesame seeds contain lysine, which when incorporated into bread improves the nutritional quality of the proteins due to a better balance of amino acids, causing a substantial increase in their utilisation by the body (FIRMINO et al., 2000). Consumers' interest in disease prevention and the search for healthier products have led the food industry to invest in products in this direction and to carry out research into functional foods, which have gained prominence due to the beneficial effects they promote for health (GÓES & PEREIRA, 2010).

2.4 Biscuit

Biscuits or biscuits are products obtained by mixing flour(s) and/or starch(es) with other ingredients, subjected to kneading and cooking processes (BRASIL, 2005). Biscuits are easy to carry and store, they complement meals and are a source of energy. They are not a staple food like bread, but are accepted and consumed by people of all ages (MORAES et al., 2010).

Wholemeal biscuits are gaining more and more market share due to their nutritional characteristics and the great current appeal for improving diet quality (Zuniga et al., 2011).

Enriching the biscuit with sesame flour is an alternative mainly due to the fact that this seed has high nutritional value, noble substances that act to prevent diseases, as well as being a crop adapted to the semi-arid region that allows for the generation of employment and income in the countryside and the movement of capital in the municipalities where it is produced and commercialised (FIRMINO et al., 2005).Oliveira; Nabeshima; Clerici, (2014) when sensorially and technologically evaluating biscuits containing defatted sesame flour and resistant starch, observed that the use of these ingredients can contribute to improving the beneficial effects in bakery products. Mosmann (2012), when making a gluten-free savoury biscuit with fibres, found that it could be an alternative to existing products on the market, as well as being nutritious and tasty.

2.5 Characteristics sensory

Sensory analysis is carried out according to the responses transmitted by individuals to the various sensations that originate from physiological reactions and are the result of certain stimuli, generating an interpretation of the intrinsic properties of the products; for this to happen there needs to be contact and interaction between the parties, individuals and products (BRASIL, 2005).

ARAÚJO et al. (2000) state that sensory characteristics stimulate the senses and provoke various degrees of desire or rejection through a complex process; consumers choose a food based on its level of sensory quality; however, sensory evaluation is the parameter that determines the rejection of a given food by the consumer, which is why its use in determining shelf life is very important (LIMA; SILVA; GONÇALVES, 1999).In the food industry, the use of modern sensory analysis techniques has been a reliable means of characterising differences and similarities in products competing for the same consumer market; optimising food attributes of appearance, aroma, flavour and texture in line with consumer market expectations; evaluating sensory changes that occur as a result of time and storage conditions, the type of packaging, variations in processing and variations in the raw material (MINIM & DANTAS, 2004).

According to Palczak et al. (2019), sensory analysis provides fundamental indications for the production and commercialisation of products, in terms of consumer preferences and demands, as well as playing an important role in the development of new products.Sensory analysis is an important parameter for evaluating food products and has been used in several studies that work on the development of foods from agro-industrial waste, such as: wholemeal bread enriched with flour from papaya by-products (SANTOS et al., 2018), loaf of bread made with graviola residue (LIMA et al., 2019), biscuits made from carrot, beetroot and wheat flour (GOUVEA et al., 2021), biscuits made with residual pequi pulp cake (DE SOUSA, al.; 2021), French bread with added vegetable flours (GHENO, et. al., 2022).

3. MATERIAL AND METHODS

3.1 Raw materials

This research used sesame seeds of the BRS SEDA cultivar, supplied by EMBRAPA - Cotton, located in Campina Grande - PB. To make the bread, the following ingredients were used: sesame residue flour (FRG) obtained from the oil extraction process, and the following products were purchased from shops in the municipality of Campina Grande: wheat flour, powdered milk, salt, margarine, sugar and biological yeast. To make the biscuits, the following ingredients were used: sesame seeds, sesame residue flour (FRG), and the other products bought in the shops, such as wheat flour, salt, oil and seasoning.

3.2 Obtaining sesame residue flour (FRG)

Sesame residue flour (SRF) was obtained from the residue (cake) generated during the sesame seed oil extraction process. The cake resulting from the oil extraction was homogenised and dried in an oven for 5 hours at a temperature of 50°C. The dried product was cooled to room temperature and ground in a blender, after which it was sieved. The flour obtained was packed in plastic bags and stored in a dry place at room temperature until the chemical, physical and physical-chemical analyses were carried out and used in the formulation of bread and biscuits.

Figure 1 - Sesame cake obtained after oil extraction (A), sesame residue flour (B) (by Authors)

3.3 Chemical, physical and physico-chemical characterisation of sesame flour

Chemical, physical and physico-chemical analyses were carried out to determine the characteristics of the sesame flour (FRG) and the processed flour products (bread and biscuits).

3.3.1. Content of water

The water content was determined using the method described by the Adolfo Lutz Institute (2008), expressed as a percentage (%).

3.3.2. Lipids

The lipid content was determined according to the methodology of Bligh and Dyer (1959).

3.3.3 pH

The pH was determined using the potentiometric method with a peagometer, previously calibrated with buffer solutions of pH 4.0 and 7.0; the results were expressed in pH units.

3.3.4 Acidity titratable

Titratable acidity was determined using the acidimetric method of the Adolfo Lutz Institute (2008), the samples of which were titrated with a standardised 0.1N NaOH solution.

3.3.5 Colour

The colour parameter was determined using a Hunter colorimeter, model Hunterlab Miniscan, with L being the luminosity, a defined as the transition from green (-a) to red (+a) and b representing the transition from blue (-b) to yellow (+b). The readings were taken in triplicate and the average values of L, a* and b* were obtained.

3.3.6 Activity from water

Water activity was determined, in triplicate, directly in a Decagon model Aqualab lite electronic meter at constant temperature ($25.0 \pm 0.30°C$).

3.4 Making bread and biscuits

3.4.1 Formulation of breads

Four formulations were used to prepare the bread: the standard formulation (control), produced with wheat flour and 0% sesame residue flour (SRF), and the formulations with wheat flour and SRF incorporated in proportions of 5%, 10% and 15% (Figure 2), as shown in Tables 1 and 2.

Figure 2 - Breads formulated with different concentrations of sesame residue flour (by Authors)

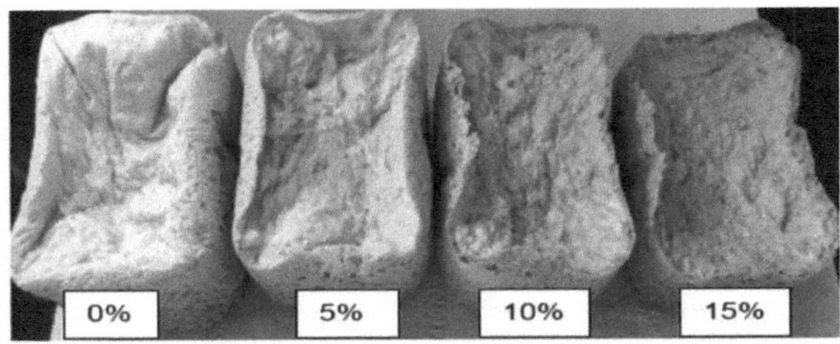

Table 1 - Proportion of wheat flour (g) and sesame residue flour (SRF) (g) used to make the bread (by Authors)

Items			FRG	
	0 %	5%	10%	15%
Wheat flour (g)	400	380	360	340
FRG (g)	0	20	40	60

In addition to wheat flour and FRG, the ingredients listed in Table 2 were used in all the formulations in fixed proportions.

Table 2 - Proportion of ingredients added to wheat flour and sesame residue flour (SRF) formulations used to make bread (by Authors)

Ingredients	Quantity
Water (ml)	200
Sugar (g)	25,0
Salt (g)	6,0
Dry biological yeast (g)	7,0
Milk powder (g)	8,0
Margarine (g)	11,0

3.4.2 Preparing bread

The ingredients were added in the following order: initially the liquid ingredients, then the dry ingredients, with the dry biological yeast being the last ingredient added to the bread machine, a Britânia Multi Pane bread machine model. Next, the cycle was chosen, in this case the type chosen was wholemeal bread, then the amount of dough was selected, where the proportion used was for loaves of around 600g; the colour of the bread crust was also selected, with the light option being used in this work. Then the bread machine was started up, where all the necessary steps in the bread-making process were carried out exclusively by the Multi Pane Bakery, during the time period pre-set by the machine of 3 hours and 30 minutes. After baking, the loaves were cooled to room temperature, sliced and placed in plastic containers until the sensory and physico-chemical analyses were carried out.

3.4.3 Formulation of biscuits

Four formulations were also used to prepare the biscuits: the standard formulation (control), produced with wheat flour only, and the formulations with wheat flour and FRG incorporated in proportions of 5, 10 and 15 per cent, as shown in Tables 3 and 4.

Table 3 - Proportion of wheat flour (g) and sesame residue flour (SRF) (g) used to make the biscuits (by Authors)

Items			FRG	
	0%	5%	10%	15%
Wheat flour (g)	300	285	270	255
FRG (g)	0	15	30	45

For all the formulations, the ingredients described in Table 4 were used in fixed proportions.

Table 4 - Proportion of ingredients added to the wheat and sesame flour (FRG) formulations used to make the biscuits (by Authors)

Ingredients	Quantity
Water (ml)	100,0
Oil (ml)	100,0
Salt (g)	3,0
Seasoning (g)	11,0
Sesame seeds (g)	27,0

3.4.4 Preparing the biscuits

First, the ingredients water, oil and salt were mixed together, then the other ingredients were added until a homogeneous dough was formed (Figure 3A). The dough was then moulded to the desired thickness of around 0.5 cm, after which it was cut to form square biscuits measuring approximately 3.0 cm. The biscuits were baked in an oven with a temperature of approximately 200°C for around 20 minutes (Figure 3B).

Figure 3 - Biscuit production stages: homogenised dough (A); baked biscuits (B) (by the authors)

3.5 Chemical, physical and physico-chemical characterisation of bread and biscuit samples

The analyses of water content, lipids, pH, titratable acidity, colour and water activity were determined according to the methodology described above.

3.6 Texture

3.6.1 Resistance to compression

The tests were carried out on a STABLE MICRO SYSTEMS model TA-TXplus texturometer. The method used consists of double compression of the sample, generating a force-time and force-distance graph, from which the values needed to calculate the texture parameters are obtained. The P/36R probe was used to measure the force (N) sufficient to compress the bread and biscuit samples (Figure 4). The samples were fixed on the base of the equipment, then a perpendicular force was applied through the cylindrical probe to the surface of the samples. The texture parameters were calculated using the curve as follows:

■Firmness: peak force measured during the first compression cycle (N);

■Cohesiveness: ratio between the areas of the second and first compression from the initial point to the peak (dimensionless);

■Elasticity: distance from the starting point of the second compression to the peak (m);

■Chewiness: product of firmness, cohesiveness and elasticity (J).

Figure 4 - Compressive strength test of bread and biscuit samples, carried out using the TA-TXplus Texturometer (by Authors)

3.6.2 Resistance to breakage

The HDP/3PB probe was used to measure the force (N) sufficient to cause the samples to break. The samples were fixed on the base of the equipment, then a perpendicular force was applied through the probe to the surface of the bread and biscuit samples. Six samples were used for each formulation and the average was obtained at the end.

Figure 5 - Breaking strength test of bread and biscuit samples, carried out using the TA-TXplus Texturometer (by Authors)

3.7 Analysing sensory

The sensory acceptance test of the breads and biscuits made with different concentrations of sesame residue flour was carried out in the Sesame

Laboratory. Food Engineering at the Federal University of Campina Grande. The acceptance test was carried out individually. Before carrying out the sensory analysis, the judges were informed about the purpose of the study in question and about the raw material, after which they signed the Free Consent Form. There were 50 untrained judges, of both sexes, aged between 18 and 60, who received a sample of each bread formulation in individual containers identified by three digits. The samples were accepted using a nine-point structured hedonic scale, which assessed the following attributes: colour, appearance, aroma, texture and flavour, with the expressions "I disliked it extremely" and "I liked it extremely" at the ends. Purchase intention was assessed using a 5-point scale with the expressions "would always eat" and "would never eat" at the ends, in which the judges indicated how much they liked or disliked each sample and purchase intention was assessed using the attitude scale according to the methodology proposed by Stone & Sidel (1993).

3.8 Analysing statistics

The experimental data on the chemical, physical and physico-chemical characterisation of the sesame residue flour (FRG) and the products (bread and biscuits) obtained from the flour were subjected to a completely randomised design and the means were compared using the Tukey test at 5% probability level, using the Assistat software, version 7.5 beta (SILVA & AZEVEDO, 2010).The results obtained from the sensory analyses were analysed using the CONSENSOR 1.1 software (SILVA; DUARTE; CAVALCANTI MATA, 2010), used for calculate the percentage of agreement between sensory analysis judges and the ASSISTAT programme, version 7.5 beta (SILVA & AZEVEDO, 2010), in which the analysis of variance was carried out and the means were compared using the Tukey test at 5% probability.

4. RESULTS AND DISCUSSION

Chemical, physical and physico-chemical characterisation of sesame residue flour (FRG)

The average values determined in the chemical, physical and physical-chemical characterisation of sesame residue flour (FRG) are shown in Table 5.

Table 5 - Chemical, physical and physico-chemical characteristics of sesame residue flour (FRG) (by Authors)

Parameters	Mean and Standard Deviation
Water content (%)	$7,1 \pm 0,10$
Lipids	$32,46 \pm 0,30$
pH	$6,17 \pm 0,01$
Titratable acidity (%)	$1,13 \pm 0,01$
Colour (L)	$62,57 \pm 0,29$
Colour (a)	$4,19 \pm 0,15$
Colour (b)	$22,24 \pm 0,54$
Aw	$0,625 \pm 0,0006$

The water content of sesame residue flour is within the limit recommended by Brazilian legislation through Resolution RDC no. 263 (BRASIL, 2005), which stipulates that flours, cereal starch and bran should have a maximum water content of 15.0 %. In this study, the average water content determined for sesame flour was 7.1%, a value considered satisfactory for maintaining the flour's quality and shelf life. This value was similar to that found by Maia et al. (1999) when studying the efficiency of sesame residue flour as a protein supplement to extruded caupi flour, at 7.21%; and higher than the 5.19% content reported by Clerici; Oliveira; Nabeshima (2013) when assessing the quality of cookie-type biscuits made by partially replacing wheat flour with defatted sesame flour.It can be seen that the lipid content of 32.46% was considered high, given that the flour is a by-product of the oil extraction process; however, Finco et al. (2011) obtained an even higher lipid value of 39.74% for sesame flour in their study on yoghurt with added sesame flour. Clerici; Oliveira; Nabeshima (2013) obtained a lipid content of 24.06% for GFD, which is lower than that reported in this study. The pH value determined for sesame flour in this study was 6.17; a similar result was reported by Amorim; Sousa; Souza (2012) who obtained a pH of 6.22 for pumpkin seed flour. The pH is a factor of great importance in limiting the ability of microorganisms to develop in food and helps to define technological procedures for preservation (SOUZA et al., 2008). Acidity represents the state of preservation of flour, involving both chemical and microbiological aspects, as microbial growth involves the production of organic acids and hydrolysis of proteins and carbohydrates (ORTOLAN; HECKTHEUER; MIRANDA, 2010). It can be seen that the average acidity value of the flour of 1.13% was lower than the limit established by legislation (BRASIL, 2005) for defatted soya flour of 2.0%.The colour parameters (L, a, b) for the sesame flour were 62.57, 4.19 and 22.24, respectively. These parameter values were lower than those obtained by Clerici; Oliveira; Nabeshima (2013) when making cookies with the partial replacement of wheat flour with defatted sesame flour, which were 67.43, 5.38 and 17.20, respectively. Thus, the sesame flour obtained in this study tends to have a darker colour compared to the FDG determined by Clerici; Oliveira; Nabeshima (2013).According to Gava; Silva; Farias (2008), Aw ranges from 0 to 1, with a value of 0.6 being considered the minimum limit for the development of microorganisms in food. The Aw value of sesame flour, equal to 0.625, is within the minimum limit for the development of microorganisms. The result obtained was lower than the value determined by Gomes; Reis; Silva (2012) for caupi bean flour, whose Aw was around 0.7; according to the authors, the product can be considered microbiologically stable, since most fungi develop in water activity above 0.8.

Chemical, physical, physico-chemical and texture characterisation of the breads

Table 6 shows the results obtained from the analyses of the physical and physico-chemical characteristics of the bread made with concentrations of 0 (standard), 5, 10 and 15% FRG sesame residue flour.

Table 6 - Chemical, physical and physico-chemical characteristics of standard bread and bread with sesame residue flour (SRF) at different concentrations (by Authors)

FRG content				
Parameters	Standard	5%	10%	15%
Water content (%)	41,39 a	36,43 b	39,32 a	39,47 a
Lipids (%)	1,90 c	3,26 b	3,53 b	5,70 a
pH	5.60 ab	5,54 b	5,65 a	5,66 a
Titratable acidity (%)	0,190 b	0,238 a	0,256 a	0,251 a
Colour (L*)	69,23 a	66,00 b	63,34 b	59,25 c
Colour (a*)	1,76 d	2,59 c	3,28 b	4,42 a
Colour (b*)	23,03 a	19.98 bc	21,48 b	18,77 c
Aw	0,959 a	0,956 a	0,959 a	0,960 a

*Means followed by the same letters in the lines do not differ by Tukey's test at 5% probability.

The water content showed a significant difference only for the bread formulated with 5% GFR. All the breads formulated with GFR had a water content of between 36 and 39%, which was lower than that determined for the standard bread, where a higher water content of 41.39% was found, these values being above the maximum limit of 35% required by Brazilian legislation (Brasil, 2005). Moreira (2007) obtained water contents of around 50% for gluten-free bread made with rice and soya flour. Moreno et al. (2020), when assessing the effects of applying resistant starch in the manufacture of bread rolls, reported values of between 37 and 40.91% for the water content of bread rolls. In the bakery sector, there is a tendency to place products with a higher water content on the market, increasing softness and giving the bread a fresher appearance. Non-compliance in terms of moisture does not pose a health risk to consumers, but it does increase the risk of contamination by moulds (a medium conducive to the proliferation of microorganisms) (Ferreira et al., 2001). In terms of lipid content, only the standard loaf and the loaf made with 15% GFR differed from the others; the lipid content of the standard loaf was lower than the values obtained for the loaves made with GFR; it can also be seen that the lipid content increased as a result of the increase in GFR in the formulation of the loaves; this was to be expected, since the lipid content of sesame flour is considered high. Silva (2015), when studying the use of sesame residue flour in food preparation, reported the flour's lipid content to be approximately 32%, as the high content of the flour results from the surplus oil after the extraction process. The average pH value of the breads varied from 5.54 to 5.66, there was an increase in this parameter as a result of the increase in FRG concentration, however, the sample formulated with 5% FRG was statistically inferior to the standard sample. A similar result was reported by Wanderley et al. (2018), who carried out the physicochemical characterisation of French bread enriched with sesame flour at concentrations of 10, 15 and 20%, and found that the incorporation of flour directly influenced the increase in the value of this parameter. Oliveira et al. (2011) also found an increase in pH due to an increase in the concentration of calcium carbonate in the bread formulation. The average acidity ranged from 0.190 to 0.256. There was no significant difference between the FRG concentrations, only the standard loaf differed statistically. The acidity content obtained for the breads formulated with GFR meets the established standards, being close to the value considered optimal (0.25% acetic acid/100 g) for this parameter. Santos et al. (2018) reported that there was an increase in the acidity index of wholemeal bread when incorporating papaya by-product flour. The average values of the L* colour parameter of the breads studied decreased as a result of the increase in FRG concentrations, this effect being more pronounced in the bread formulated with 15% FRG, the result of which indicates that there was a darkening of the breads as a result of the increase in formulations. According to Purlis (2011), breads with a luminosity of around 70 have good sensory acceptance. However, values below 60 result in excessive darkening and above 78 in a very light colour, indicating insufficient baking. With regard to the a* parameter, the breads formulated with FRG showed higher values than the standard bread, with a shift in the colour of the samples with FRG towards red compared to the standard sample. For the b* parameter, there was a significant difference between the samples. The values for this parameter ranged from 18.77 to 23.03, with the maximum value being obtained for the standard loaf. Feitosa et al. (2013) assessed the colour of the crumb of French bread and obtained an average value of 14.3 for the b* parameter. Aw is directly related to the preservation of food, as well as being related to a faster or slower ageing of the bread (Ditchfield, 2000). The average Aw value ranged from 0.956 to 0.960, with no significant difference between the formulations evaluated. Bread is classified as a food with Aw between 0.98 and 0.93. These Aw values inhibit the growth of salmonella, Cl.botulinum and other pathogenic bacteria; the effect is accentuated when unfavourable conditions coexist with other agents, such as pH and temperature (Ordóñez, 2005). Moreno et al. (2020) obtained Aw values above 0.93 for bread made with resistant starch. Pires et al. (2018) analysed making gluten-free and vegetable flour-based breads, also reported Aw values higher than 0.9. The results for the texture parameters (firmness, cohesiveness, elasticity, chewiness and resistance to breakage) of the breads formulated with different concentrations of sesame residue flour are shown in Table 7.

Table 7 - Texture parameter measurements (firmness, cohesiveness, elasticity, chewiness and resistance to breakage) of breads formulated with different concentrations of sesame residue flour (by Authors)

Parameters					
Treatments	Firmnes s (N)	Cohesiven ess	Elasticity	Chewability (N)	Resistance to breakage
Standard	2,84 d	0,83 a	0,81 a	1,21 d	13,20 a
5%	4,93 c	0,77 b	0,73 b	2,33 c	13,41 a
10%	7,80 b	0,69 c	0,65 c	3,43 b	12,13 a
15%	11,02 a	0,67 c	0,64 c	4,54 a	9,53 a
DMS	1,63	0,04	0,05	0,72	7,00

*Means followed by the same letters in the columns do not differ by Tukey's test at 5% probability.

The firmness parameter was influenced by the addition of sesame residue flour to the bread formulation, with an increase in this parameter as a result of the addition of the flour. The increase in firmness is probably related to the high fibre content characteristic of this flour, making the bread have a firmer texture. Borges et al. (2013) also found an increase in this parameter when they increased the incorporation of quinoa flour as a partial substitute for wheat flour when making bread.With regard to cohesiveness, which represents the maximum extent to which a material can be deformed before breaking, it can be seen that there was a reduction in this parameter as a result of increasing the concentrations of sesame residue flour in the formulation of the breads. It was found that the cohesiveness averages ranged from 0.83 to 0.67, with no statistical difference between the breads formulated with concentrations of 10 and 15% FRG. Elasticity was also reduced with the increase in the incorporation o f sesame residue flour, in which case the speed required for the deformed product to return to its initial state was reduced with the increase in GFR.

Evangelho et al. (2012) found that increasing the percentage of rice flour in the bread formulation and increasing the storage time led to a reduction in the elasticity of the bread. The chewiness parameter represents the energy required to chew the bread. In this study, the addition of sesame residue flour at 5, 10 and 15 per cent contractions interfered with the chewiness of the bread, promoting an increase of 48.06, 64.72 and 73.35 per cent in relation to standard bread, respectively. Silva et al. (2009) developed a loaf of bread with the addition of okara flour and found that the increase in chewiness was influenced by the addition of okara flour as a partial substitute for wheat flour and also by the storage period of the loaves. As for the resistance to breakage parameter, the treatments evaluated did not differ statistically. However, the increase in sesame residue flour in the bread formulation led to a decrease in resistance, with the highest resistance index being obtained for the standard loaf.

Sensory analysis of the bread

The results of the sensory analysis of loaves formulated with different concentrations of sesame residue flour in terms of colour, appearance, aroma, texture and flavour are shown in Table 8. The results showed that the standard sample obtained higher averages than the samples formulated with sesame residue flour for most of the attributes assessed, with the exception of the texture parameter, which was higher for the sample formulated with 5% FRG.

Table 8 - Results relating to the means and coefficients of agreement of the sensory analysis of the bread formulations (by Authors)

PARAMETERS									
TREAT.	Colour		Appearance		Flavour		Texture		Flavour	
	Average	CC(%)	Average	CC(%)	Average	CC(%)	Average	CC(%)	Average	CC(%)
0%	7,56a	39,36	7,70a	41,15	7,18a	37,96	7,22a	38,32	7,36a	38,43
5%	7,26a	39,47	7,48a	42,23	6,74ab	29,72	7,30a	39,24	7,28a	40,60
10%	7,08ab	33,96	7,02a	37,60	6,22b	24,21	6,64ab	34,09	5,82b	22,47
15%	6,44b	27,19	6,24b	24,76	6,08b	22,45	6,32b	33,56	5,54b	23,83

*Means followed by the same letters in the columns do not differ by Tukey's test at 5% probability.

The standard and 5% samples did not differ statistically for all the attributes assessed, and the highest level of acceptance was obtained by these samples. Ferreira et al. (2021) carried out a sensory analysis of loaves enriched with jaboticaba peel flour and found that all the attributes assessed in the acceptability test obtained decreasing scores as a result of the increase in flour concentration. The authors reported that, in general, the breads with no added flour and 5 per cent added flour were the most acceptable to the judges. Gurjão et al. (2014), when assessing the sensory properties and market acceptance of bread enriched with pumpkin grain flour (PGF), also found that the standard and 5% PGF treatments were more acceptable than the 10 and 15% PGF concentrations.With regard to the colour parameter, it could be seen that the treatment with the highest agreement was the bread formulated with 5%, with an average score of 7.26 and an agreement coefficient of 39.47%, followed by the standard sample, with an agreement coefficient of 39.36%. It is clear that the The incorporation of flour resulted in a darker product, causing the judges to reject it more. For the sensory parameter appearance, the highest coefficient of agreement (42.23%) was obtained for the bread formulated with 5% FRG. With regard to the aroma parameter, the highest average score was obtained for the standard loaf, with an average value of 7.18 and a coefficient of agreement of 37.96%. As for the texture parameter, the highest average score and the best coefficient of agreement were obtained for the bread with 5% added, with values of 7.30 and 39.24% respectively.

The flavour sensory parameter obtained the highest average for the standard loaf, with a value of 7.36. When evaluating the best coefficient of agreement, it was noted that the bread formulated with 5% had the highest value (40.60%), for an average of 7.28. It can be inferred that the more pronounced aftertaste in the 10 and 15% FRG concentrations had a negative influence on the acceptability of the samples, resulting in the lowest scores given by the judges among the sensory parameters assessed. Stikic et al. (2012), when assessing the sensory quality of breads containing quinoa, found that concentrations higher than 15% altered the flavour, making the bread slightly bitter.

In general, it was observed that the bread formulated with 5% FRG obtained the best coefficients of agreement, along with the control bread formulated without the addition of FRG. A similar result was obtained by Maia et al. (2015), who reported that the highest acceptance index was obtained for bread made with 5% added coconut flour, which was superior to bread formulated with 0 and 7.5% flour in all the attributes assessed. It was also found that increasing the concentration of sesame residue flour in the bread formulation led to a reduction in acceptability The greatest rejection was attributed to the bread formulated with 15%, with lower averages than the other treatments in all the parameters evaluated. This was also reported by Bruni et al. (2020) when they sensorially analysed cheese bread with the addition of sweet potato flour at concentrations of 0, 25, 50 and 100%, and reported that an increase in the incorporation of flour resulted in a reduction in the acceptability of the bread by the judges.

The intention to consume the loaves formulated with different concentrations of FRG is shown in figure 6. When evaluating the percentage of intention to consume the control bread sample, without the addition of sesame residue flour, it was noted that none of the tasters claimed that they would never eat it, only 4% of the tasters would rarely eat the product, 18% of the tasters would maybe eat/perhaps not eat the product, 38% would always eat it and the vast majority, 40% would often eat the standard bread without the addition of FRG. According to the results obtained, it can be said that the product was well accepted by the tasters.

Figure 6 - Intention to consume bread rolls formulated with different concentrations of sesame residue flour (by Authors)

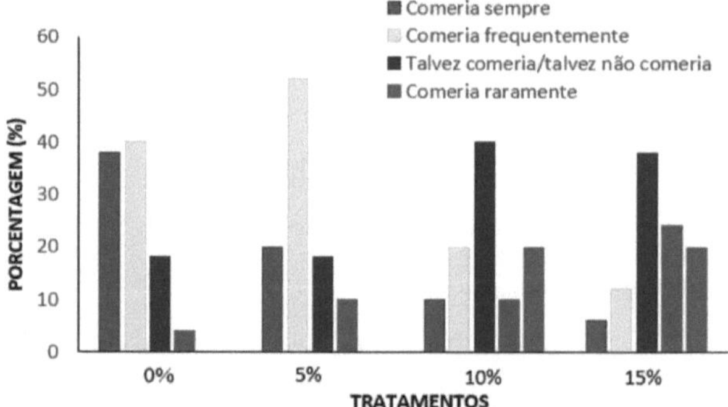

For the bread formulated with 5% FRG, good acceptance was observed by the tasters, in which the majority of the tasters, 52%, reported that they would often eat the sample, 20% claimed that they would always eat it, only 10% of the tasters would rarely eat the product, 18% of the tasters would maybe eat/perhaps not eat the product and none of the tasters claimed that they would never eat this formulation, demonstrating good acceptability of the product by consumers. Regarding the bread formulated with 10% sesame residue flour, it was found that the increase in FRG had a negative influence on the product's acceptability, so that 20 per cent said they would never eat this sample, while 40 per cent of the tasters said that they might/perhaps wouldn't eat it. Despite the fact that this sample had a considerable percentage of rejection, it can be said that the product was reasonably accepted by the judges. The intention to consume the bread formulated with 15% sesame residue flour resulted in the lowest acceptability among the other formulations evaluated, with only 6% of the judges claiming that they would always eat the bread formulated with 15% FRG. Notably, the addition of sesame residue flour to the bread formulations led to greater rejection among the judges, so that the bread formulated with 15% FRG was considered the least attractive product by the majority of consumers. This fact was also reported in studies carried out by: Santos et al., (2018) when sensorially analysing wholemeal bread enriched with papaya by-product flour and; by Khalil et al. (2017) when sensorially evaluating unleavened bread supplemented with green banana flour (FBV).

Chemical, physical, physico-chemical and texture characterisation of the biscuits

The results of the chemical, physical and physico-chemical analyses of the standard biscuit and the biscuit formulated with different concentrations of sesame residue flour (SRF) are shown in Table 9.

Table 9 - Chemical, physical and physico-chemical characteristics of standard biscuits and biscuits containing sesame residue flour (SRF) at different concentrations (by Authors)

FRG content				
Parameters	Standard	5%	10%	15%
Water content (%)	2,19 b	2,17 b	2,55 b	4,07 a
Lipids (%)	23,72 b	24,02 b	26,45 a	26,87 a
pH	5,67 a	5,70 a	5,66 a	5,65 a
Titratable acidity (%)	0,107 a	0,104 a	0,111 a	0,111 a
Colour (L*)	54,42 a	49,05 c	51,55 b	52,31 b
Colour (a*)	6,62 d	10,48 a	8,43 b	7,96 c
Colour (b*)	31,19 a	31,46 a	30.33 ab	29,34 b
Aw	0,297 b	0,298 b	0.300 ab	0,302 a

*Means followed by the same letters in the lines do not differ by Tukey's test at 5% probability.

The water content of the biscuits was influenced by the increase in FRG formulations, where the maximum value obtained for this parameter was 4.07% for the biscuit formulated with 15% FRG. According to Oliveira; Nabeshima; Clerici (2014), moisture is linked to the crunchy texture of biscuits, which is a very important property, since biscuits that absorb moisture during storage lose their crunchiness, which causes sensory damage to the product. The biscuits produced in this study are within the standards of current legislation for biscuits (BRASIL, 1978), which determines a maximum of 14 per cent. The results obtained in this study were lower than those reported by Clerici; Oliveira; Nabeshima (2013) who obtained contents of 5.96 for the standard cookie and 4.79 for the cookie made with 10% FDG. Firmino; Sila; Sousa (2005) found a content of 1.40 when making a biscuit with GFD.The biscuits made with FRG had higher lipid levels than the standard biscuit, where the levels ranged from 23.72 to 26.87, increasing as a result of the addition of sesame flour. This is due to the high lipid content of sesame residue flour. The values obtained for this parameter were similar to those found by Santos et al. (2011) for a buriti flour biscuit with added oats, in which the lipid content was 22.46%.

As for pH, the biscuits did not differ statistically depending on the formulations used. The values obtained were very close to 5.7, which is below the range considered normal for biscuits (between 6.5 and 8.0), as mentioned by Pyler (1982). It can also be seen that there was a reduction in this parameter as a result of the increase in the concentration of FRG. A similar behaviour was observed by Acorsi (2009) when they reported a reduction in the pH of biscuits as a result of an increase in the concentration of pine nut flour.

The acidity levels determined for the biscuits with different concentrations of FRG ranged from 0.104 to 0.111%, with this parameter increasing as the biscuit formulations increased. These values were within the standards set for biscuits by legislation (BRASIL, 1978), which determines a maximum acidity of 2.0 mL/100 g in normal solution. In contrast, Andrade (2013) observed a reduction in titratable acidity of 0.20, 0.17 and 0.16% citric acid when making biscuits with concentrations of 10, 20 and 30% green banana flour.

The L* parameter of the standard biscuit was higher than the values obtained for the biscuits made with different concentrations of GFR, which shows that the biscuits formulated with GFR had a greater darkening compared to the standard biscuit. The biscuits produced in this study had lower L* luminosity values than those reported by Oliveira; Nabeshima; Clerici (2014) for the cookie made with 10% FDG (65.64); however, these values were similar to those reported by Andrade (2013) in biscuits formulated with 10, 20 and 30% FDG. green banana flour, which obtained L* values of 53.66, 54.14 and 56.55, respectively.

The a* parameter values were lower for the standard biscuit compared to the biscuits formulated with FRG. The reduction in the a* parameter observed in the standard biscuit probably occurred because the biscuit was formulated with wheat flour only, making it lighter than the other biscuits.The biscuits made in this study obtained b* values higher than those reported by Perez and Germani (2007) for biscuits made with aubergine flour (b*= 18.85 to 20.73); and by Oliveira et al. (2014), when developing cookies with defatted sesame flour and resistant starch (b*= 23.25 to 24.42). It can also be seen that the biscuits formulated with 10% and 15% GFR differed statistically from the other treatments, but it can be seen that there was a predominance of yellow colour in all the biscuits evaluated.

The water activity (Aw) ranged from 0.297 to 0.302, with only the biscuits formulated with 10 and 15 per cent FRG differing statistically from the other concentrations, and this parameter increased as a result of the addition of FRG to the biscuit formulation. Baptista et al. (2012) also found an increase in Aw with the addition of Moringa oleifera in the formulation of cookies.

Table 10 shows the results of the texture parameters of the biscuits made with different concentrations of sesame residue flour.

With regard to the firmness parameter, it can be seen that the treatments evaluated did not differ statistically, which indicates that the incorporation of sesame residue flour did not cause significant changes in the firmness of the biscuits. The firmness values obtained in this study were higher than those determined by Moura et al. (2014), when they produced biscuits enriched with brown linseed flour, whose values ranged from 83 to 112 N. The harder texture in biscuits can be attributed to the increased protein content and its interaction during dough development and baking (MCWATERS et al., 2003).

Table 10 - Texture parameter measurements (firmness, cohesiveness, elasticity, chewiness and resistance to breakage) of biscuits formulated with different concentrations of sesame residue flour (by Authors)

Parameters					
Treatments	Firmness	Cohesiveness	Elasticity	Chewability	Resistance to breakage
Standard	345,63 a	0,63 a	0,52 b	126,71 b	19,36 a
5%	385,57 a	0,62 a	0,51b	124,09 b	15.48 ab
10%	405,92 a	0,68 a	0.60 ab	171,90 a	12.90 ab
15%	446,69 a	0,72 a	0,62 a	174,81 a	11,93 b
DMS	136,0	0,10	0,09	17,98	7,01

*Means followed by the same letters in the columns do not differ by Tukey's test at 5% probability.

With regard to the cohesiveness of the biscuits, it can be seen that there was no significant variation between the formulations evaluated; however, it can be seen that the increase in sesame residue flour led to an increase in this parameter. The elasticity of the biscuits also increased as a result of the addition of sesame residue flour as a partial substitute for wheat flour, with no significant difference between the biscuits formulated with 0% (standard) and 5% FRG.

The chewiness parameter was influenced by the incorporation of sesame residue flour. In this study, the energy required during the chewing process was higher for biscuits formulated with sesame residue flour, probably due to the high fibre content of this flour. As reported by several authors, the incorporation of sesame flour into biscuits has led to a significant increase in dietary fibre content: cookies (CLERICI; OLIVEIRA; NABESHIMA, 2013), rice-based biscuits (TAIKETI et al., 2010) and biscuits developed with defatted sesame flour and resistant starch (OLIVEIRA; NABESHIMA; CLERICI, 2014). The resistance to breakage of the biscuits was reduced with the addition of sesame residue flour to the formulations, and the partial substitution of wheat flour with FRG made the samples more fragile and susceptible to breakage. Moreno et al. (2010) reported in their study that the use of parboiled rice flour to replace wheat flour resulted in more fragile biscuits, due to the absence of the gluten protein network in the rice.

Sensory analysis of the biscuits

The results of the sensory analysis of the biscuits formulated with different concentrations of sesame residue flour (SRF) in terms of colour, appearance, aroma, texture and taste are shown in Table 11.

Table 11 - Acceptance evaluation averages of standard biscuit samples and those with added sesame residue flour (by Authors)

Treatment	Parameters				
	Colour	Appearance	Flavour	Texture	Flavour
Standard	7,46 a	7,58 a	7,34 a	7,56 a	7,18 a
5%	7.24 ab	7.20 ab	6,96 a	7,50 a	7,06 a
10%	6,68 b	6,84 b	6,92 a	6,94 a	6.36 ab
15%	6,68 b	6,86 b	6,66 a	6,96 a	5,92 b
DMS	0,68	0,67	0,73	0,63	0,97

*Means followed by the same letters in the columns do not differ by Tukey's test at 5% probability.

The results showed that the biscuits in the control formulation (with no added FRG) and those formulated with 5% FRG had the best average acceptability in all the parameters assessed, with average scores above 6.9, demonstrating good acceptability among the judges in terms of the attributes assessed. It was found that the biscuits formulated with 10 and 15% GFR were less acceptable to tasters, resulting in lower scores than the other samples evaluated. Silva et al. (2019), when carrying out a sensory analysis of cookie-type biscuits made from avocado pit flour, reported that increasing the concentration of flour in the biscuit formulation resulted in a reduction in the scores of all the attributes evaluated. Clerici et al. (2013) also reported that increasing the incorporation of defatted sesame flour interfered with the acceptance of the cookies by the judges, so that cookies formulated with concentration F3 (30% FDG) resulted in greater rejection of the product. In general, it was observed that the control biscuit (formulated without the addition of FRG) obtained higher averages and coefficients of agreement than the other samples evaluated, for most of the attributes evaluated, with the exception of the colour parameter, where the coefficient of agreement attributed by the judges was higher for the biscuit formulated with 5% FRG (45.31%). It was also noted that the scores given by the judges decreased as a result of the increase in the concentration of sesame residue flour. It is possible to infer that the residual flavour characteristic of this oilseed interfered with the palatability of the samples, so that the addition of GFR to the biscuit formulation contributed to a reduction in the acceptability of the products by the evaluators. Although the 10 and 15% GFR formulations resulted in lower acceptability by the judges, it was found that the flavour and texture attributes did not differ statistically between the GFR formulations evaluated. However, it is important to emphasise that sesame residue flour has great potential for acceptance, since the biscuit formulated with 5% GFR was well accepted by consumers.

These results show that this flour can be used as a partial substitute for wheat flour in food production for the purpose of protein enrichment.Miranda et al. (2021), when making gluten-free biscuits enriched with orange waste flour, also found that the addition of waste flour resulted in a reduction in the acceptability of the biscuits in all the attributes assessed. Figure 7 shows the intention to consume biscuits formulated with different concentrations of GFR.

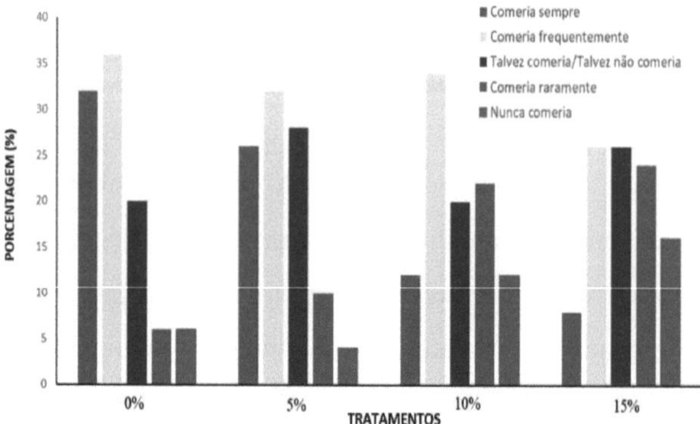

With regard to the control biscuit (0% FRG), it was found that only 6% of the tasters claimed that they would never eat it, a percentage of 6% was also attributed to the intention to eat the product rarely, 20% of the tasters would perhaps eat/perhaps not eat the product, 32% would always eat it and the vast majority, 36% would always eat it. would often eat the biscuit without added GFR. Based on the results obtained, it can be said that the product was well accepted by the judges. The biscuit formulated with 5% FRG was evaluated positively by the tasters, with 32% of the tasters saying they would often eat the sample, 26% saying they would always eat it, only 10% of the tasters saying they would rarely eat the product, 28% of the tasters saying they might eat it/wouldn't eat it and 4% of the tasters saying they would never eat this formulation, resulting in the product being well accepted by consumers.

For the biscuit formulated with 10% sesame residue flour, it was found that the increase in FRG had a negative influence on the product's acceptability, with 22% saying that they would rarely eat this sample. However, 34% of the tasters said that they might eat it/wouldn't eat it, showing that the product was reasonably accepted by the tasters, despite the fact that the sample had a considerable percentage of rejection. The biscuit formulated with 15% sesame residue flour resulted in the lowest consumption intention among the other formulations evaluated, in which only 8% of the judges claimed that they would always eat the biscuit formulated with 15% GFR. Notably, the increase in the incorporation of sesame residue flour in the biscuit formulations had a negative influence on the acceptability of the product by the judges, making the biscuit formulated with 15% FRG a less palatable product for consumers. A similar result was reported by Gaspar et al. (2020) when they made biscuits with residual flours from pumpkin, beetroot and carrot peelings, and by Gouvea et al. (2021) when they assessed the purchase intention of biscuits made from a mixture of beetroot stalk flour, oat flakes and wheat flour.

5. CONSIDERATIONS FINAL

Sesame flour can be considered a good alternative for utilising the waste generated during the oil extraction process. The parameters determined in the characterisation of the SGRF met the standards recommended by Brazilian legislation and it is considered to be a good quality product. Partially replacing wheat flour with sesame residue flour resulted in changes to the chemical, physical and physicochemical properties of the breads and biscuits, but did not cause any loss of product quality. The texture of the breads and biscuits was influenced by the incorporation of sesame residue flour, with changes occurring in most of the parameters assessed depending on the formulation used. Among the formulations developed with sesame residue flour, the formulation with 5% added FRG was the most acceptable, similar to the scores given to the bread and biscuits in the control formulation. The use of sesame residue flour to supplement breads and biscuits not only helps to develop practical food products with added nutritional value, but also makes it possible to use the residue resulting from the sesame oil extraction process properly.

6. REFERENCES

ACORSI, D. M.; BEZERRA, J. R. M. V.; BARÃO, M. Z.; RIGO, M. Viability of the processing of biscuits with pine nut flour. **Revista do Setor de Ciências Agrárias e Ambientais**, v. 5, n. 2, p. 207-212, 2009.

AMORIM, A. G.; SOUSA, T. DE A.; SOUZA, A. O. DE. Determination of pH and titratable acidity of pumpkin seed flour (cucurbita maxima). In: CONGRESSO NORTE NORDESTE DE PESQUISA E INOVAÇÃO, 7 , 2012, Palmas.

Proceedings... Palmas - TO, 2012.

ANDRADE, C. K. O. **Elaboration and acceptability of biscuits enriched with green banana flour.** 2013. 50f. Monograph (Degree in Agricultural Sciences) - State University of Paraíba, Catolé do Rocha, 2013.

A R A Ú J O , A. H.; FONTENELE, A. M. M . ; MOTA, A. P. M.; DANTAS, F. F.;VERRUMA-BERNADI, M. R. Sensory analysis of fresh coconut water compared to pasteurised water. In: BRAZILIAN CONGRESS OF FOOD SCIENCE AND TECHNOLOGY, 17, 2000, Fortaleza. **Proceedings...** Fortaleza: SBCTA, 2000. CD-ROM.

ARRIEL, N. H. C.; VIEIRA, D. J.; FIRMINO, P. T. **Current situation and perspectives of sesame cultivation in Brazil.** Genetic Resources and Plant Breeding for the Brazilian Northeast, 2006.

ASSIS, L. M. de; ZAVAREZE, E. da R . ; RADÜNZ, A. L.; DIAS, A. R. G.; GUTKOSKI,L. C.; ELIAS, M. C. Nutritional, technological and sensory properties of biscuits with wheat flour replaced by oat flour or parboiled rice flour. **Alimentos Nutrição**, Araraquara, v.20, n.1, p. 15-24, jan./mar. 2009.

BAPTISTA, A. T. A.; SILVA, M. O.; BERGAMASCO, R.; VIEIRA, A. M. S. Physico-chemical and sensory evaluation of biscuits made with Moringa oleifera leaves. **Bulletin of the Food Processing Research Centre**, Curitiba, v. 30, n. 1, p. 65-74, jan./jun. 2012

BLIGH, E.G.,DYER, W.J. A rapid method of total lipid extraction and purification. **Canadian Journal Biochemistry Physiological**, 27(8), 911-917, 1959.

BORGES, J. T. DA S.,VIDIGAL, J. G.,SILVA, N. A. DE S.,PIROZI, M. R.,PAULA,

C. D. DE. Physico-chemical and sensory characterisation of bread containing mixed wheat and quinoa flour. **Revista Brasileira de Produtos Agroindustriais**,15(3), 305-319, 2013.

BRAZIL. National Health Surveillance Agency. Resolution RDC no. 263, of 22 September 2005. **Technical regulations for cereal products, starches, flours and e bran.** Retrieved from https://bvsms.saude.gov.br/bvs/saudelegis/anvisa/2005/rdc026_22_09_2005.html, 2005.

BRAZIL. Ministry of Health. National Health Surveillance Agency. Resolution CNNPA. No. 12 of 24 July 1978. Identity and Quality Standards for Foods and Beverages. **Official Federal Gazette,** Brasília, DF, 27 July 1978.

BRAZIL. Ministry of Health. National Health Surveillance Agency. Ordinance SVS/MS no. 27, of 13 January 1998. Technical Regulation on Complementary Nutritional Information. **Diário Oficial da União,** Brasília, DF, 16 Jan 1998.

BORNEO, R; AGUIRRE, A. Chemical composition, cooking quality, and consumer acceptance of pasta made with dried amaranth leaves flour. **LWT - Food Science and Technology**, v. 41, p. 1748-1751, 2008.

BRUNI, A. R. da S.; MORETO, V. O.; CZAIKOSKI, A.; CZAIKOSKI, K. Development and sensory analysis of cheese bread with addition of sweet potato flour. **Brazilian Journal of Development**, [S. l.], v. 6, n. 8, p. 58391-58403, 2020.

CLERICI, M.T.P.S.; OLIVEIRA, M.E. de; NABESHIMA, E. H. Physical quality,chemical and sensory characteristics of biscuits made by partially replacing wheat flour with defatted sesame flour. **Brazilian Journal of Food**

Technology, Campinas, v. 16, n. 2, p. 139-146, Apr./Jun. 2013.

DE SOUSA, E. O.; MOREIRA DOS SANTOS, A. M.; DA SILVA DUARTE, A. M.;

GONÇALVES DA SILVA, M. T. Use of pequi pulp residual pie flour (caryocar coriacium wittm) in the development and characterisation of sequilho crackerv. **Revista Brasileira de Engenharia de Biossistemas**, Tup, v. 15, n. 4, p. 632-643, 2021.

DIETCHFIELD, C. **Study of methods for measuring water activity.** 195f. Dissertation (Master's in Chemical Engineering). USP Polytechnic School, São Paulo, 2000.

EL-DASH, A. A.; GERMANI, R. **Tecnologia de farinha mistas: uso de farinha mista na produção de biscoitos.** 1. ed. Empresa Brasileira de Pesquisa de Tecnologia Agroindustrial de Alimentos (EMBRAPA - SPI): Brasília, 1994, 47p.

EMBRAPA COTTON. **Sesame cultivation: product presentation**. Campina Grande, 2006. Available at: http://sistemasdeproducao.cnptia.embrapa.br. /HTMLSources/. Gergelim/ Cultivodo Sergelim/ composicaoquimica.html. Accessed on: 21 April 2014.

GOSPEL, J. A. DO,PINTO, V. Z . ,ZAVAREZE, E. DA R.,VANIER, N. L.,DIAS, A. R. G.,BARBOSA, L. M. P. Technological and nutritional properties of breads prepared with different proportions of rice flour and extruded rice flour. **Revista Brasileira de Agrociência**, 18(4), 264-282, 2012.

FEITOSA, L. R. G. DE F.,MACIEL, J. F.,BARRETO, T. A.,MOREIRA, R. T. Quality evaluation of French bread by instrumental and sensory methods. **Semina: Agricultural Sciences**, 34(2),693-704, 2013.

FERREIRA, S. M. R., OLIVEIRA, P V. DE,PRETTO, D. Quality parameters of French bread. **B.CEPPA**, 19(2),301-318, 2001.

FERREIRA, S. P. L.; JARDIM, F. B. B.; DA FONSECA, C. R.; COSTA, L. L.; ALVES, L. Sensory analysis of wholemeal bread rolls enriched with jaboticaba peel flour. **Brazilian Journal of Science, Technology and Innovation**, Uberaba - MG, v. 6, n. 1, p. 1-12, 2021.

FIGUEIREDO, A.S.; MODESTO FILHO, J. Efeito do uso da farinha desengordurada do Sesamum indicum L. nos níveis glicêmicos em diabéticas tipo 2. **Revista Brasileira de Farmacognosia**, v.18, p.77-83, 2008.

FINCO, A. M. DE O.; GARMUS, T. T.; BEZERRA, J. R. M. V.; CÓRDOVA, K. R.V. Preparation of yoghurt with added sesame flour. **Revista Ambiência.** v. 7, n. 2, p. 217-227, May/Aug. 2011.

FIRMINO, P. de T. GERGELIM: production systems and their verticalisation process, aimed at productivity in the field and improving the quality of human food, Campina Grande: **Embrapa-CNPA**, 1996. (Young Scientist Award).

FIRMINO, P. de T.; ARRIEL, N. H. C.; PEREIRA, J. R.; SILVA, M. B. da; ALMEIDA, V. DE; SANTOS, E. G. **Utilisation of Sesame (Sesamum indicum L) in Bakery Products** (Technical Communication) 2000. Campina Grande, PB Ministry of Agriculture, Livestock and Supply.

FIRMINO, P. de T.; SILVA, A. C.; SOUSA, M. E. R. de. **Sesame biscuit: nutritional quality in school meals.** Campina Grande: Embrapa Cotton, 2005.

GASPAR, P. B.; SPOTO, M. H. F.; BORGES, M. T. M. R.; BERNARDI, M. R. V.

Elaboration of flours and biscuits with residues from the family agroindustry / Elaboration of flours and cookies with residues from the family agroindustry. **Brazilian Journal of Development**, [S. l.], v. 6, n. 5, p. 25488-25506, 2020.

GAVA, A. J.; SILVA, C.A.B. da; FRIAS, J. G. **Tecnologia de alimentos: Principles and applications.** p. 95-98, São Paulo: Nobel, 2008.

GHENO, A. M., GEADICKE, J. P., MÜLLER, L., STOFFEL, F., & BARBOSA, R.

G. Evaluation of technological attributes of French bread with added vegetable flours. **Brazilian Journal of Food Technology**, 25, e2021113, 2022.

GÓES, M.S.; PEREIRA, C.A.M. Functional properties of linseed. **Nutrição Brasil Magazine.** Year 9, n. 2, p. 132-140, 2010.

GOMES, G. M. S.; REIS, R. C.; SILVA, C. A. D. T. da. Obtaining flour from cowpeas (vigna unguiculata l. walp). **Revista Brasileira de Produtos Agroindustriais**, Campina Grande, v.14, n.1, p.31-36, 2012.

GOUVEA, I. F. S. .; MACIEL, M. P. R. .; CARVALHO, E. E. N. .; CIRILLO, M. Ângelo .; BOAS, B. M. V. .; NACHTIGALL, A. M. . Proximate composition and purchase intention of biscuits obtained by mixing the flours of beet stalks, oat flakes and wheat flour. **Research, Society and Development**, [S. l.], v. 10, n. 5, p. e21110514813, 2021.

GURJÃO, F. F.; QUEIROZ, A. J. DE M.; FIGUEIRÊDO, R. M. F. de; PESSOA, T.;CARNEIRO, G. G; PAIVA, K. M. R. Analysis of the sensory properties and market acceptance of bread enriched with pumpkin grain flour. **Tecnologia & Ciência Agropecuária**, João Pessoa, v.8, n.1, p.1-3, 2014.

ADOLFO LUTZ INSTITUTE (2008). **Chemical and physical methods for analysing food.** Analytical standards of the Adolfo Lutz Institute. São Paulo, 1020p.

KHALIL, M .; TABIKHA, M .; HOSNY, M.; KORTAM, A. Physiochemical and Sensory Evaluation of some Bakery Products Supplemented with Unripe Banana Flour as a Source of Resistant Starch. **Journal of Food and Dairy Sciences**, v. 8, n. 10, p. 411-417, 2017.

LIMA, D. V.; AZEVEDO, O. O. da C.; SILVA, N. de S.; SILVA, G. S.; PONTES, E. D. S.; ARAUJO, M. G. G. de; PEREIRA, D. E.; PIOVESAN, N.; MEDEIROS, R. G.; SOARES, J. K. B.; VIERA, V. B. Development and sensory evaluation of added bread from soursop residue flour. **Research, Society and Development**, [S. l.], v. 9,n. 1, p. e172911857, 2020.

LIMA, J. R.; SILVA, M. A. A. P.; GONÇALVES, L. A. G. Sensory characterisation of fried and salted cashew nut kernels. **Ciência e Tecnologia de alimentos**, Campinas, v. 19, n 1, p. 123-126. 1999.

MAIA, G. A.; CALVETE, Y. M. A.; TELLES, F. J. S.; MONTEIRO, J. C. S.; SALES,M. G. Efficiency of defatted sesame flour as a protein supplement to extruded caupi flour. **Pesquisa Agropecuária Brasileira**, Brasília, v. 34, n. 7, p. 1295-1303, 1999.

MAIA, J. D.; BARROS, M. de O.; CUNHA, V. C. M.; SANTOS, G. R. dos; CONSTANT, P. B. L. Acceptability study of bread enriched with coconut pulp residue flour. **Revista Brasileira de Produtos Agroindustriais**, Campina Grande, v.17, n.1, p.1-9, 2015.

MCWATTERS, K. H.; OUEDRAOGO, J. B.; RESURRECCION, A. V. A.; HUNG,Y. C.; PHILLIPS, R. D. Physical and sensory characteristics of sugar cookies containing a mixture of fonio (Digitaria exilis) and cowpea (Vigina unguiculata) flours. **International Journal of Food Science and Technology**, v. 38, n. 4, p. 403-410, 2003.

MILANI. M.; GONDIM, T. M. S.; COUTINHO, D. **Cultura do sergelim,** Technical Circular 83, Campina Grande, p. 1-10, 2005.

MINIM, V. P. R.; DANTAS, M. I. S. Sensory evaluation of minimally processed products. In: NATIONAL MEETING ON MINIMAL PROCESSING OF FRUIT AND VEGETABLES, 3, 2004, Viçosa. **Proceedings...** Viçosa: UFV, 2004. p. 33-37, 2004.

MMA - MINISTRY OF THE ENVIRONMENT. **Identification of Technological Alternatives for the Control, Treatment and Reuse of Industrial Waste.** 2006. Available at: http://www.mma.gov.br/estruturas/sqa_pnla/_arquivos/item_8.pdf. Accessed on:

12/01/2015.

MIRANDA, M. S.; BARROS, V. C.; HUNALDO, V. K. L.; SANTOS, L. H. dos ; FREITAS, A. C. de .; LOBATO, J. S. M.; FONTENELE, M. A. . Gluten-free cookies enriched with orange residue flour. **Research, Society and Development**,[S. l.], v. 10,n. 11, p. e64101119023, 2021.

MORAES, K. S.; ZAVAREZE, E. R.; MIRANDA, M. Z.; SALAS-MELLADO, M.

M. Technological evaluation of cookie-type biscuits with variations in lipid and sugar content. **Ciência e Tecnologia de Alimentos**, Campinas, v. 30, p. 233-242, 2010.

MOREIRA, M. R. (2007). **Preparation of a premix for gluten-free bread for coeliacs**. 104 f. Dissertation (Master's) - Postgraduate Programme in Food Science and Technology, Federal University of Santa Maria, 2007.

MOREIRA, D. K. T.; BARCELOS, M. F. P. **Partial utilisation of extruded sesame flour in rice-based biscuits.** (Technical Communication) 2010. Rio de Janeiro: Embrapa.

MORENO, H. O.; DELLANOCE, P. GUASTAFERRO,E. A. **Effects of applying of resistant starch in the manufacture of bread rolls**. Mauá School of Engineering - Mauá Institute of Technology, 2010.

MOSSMANN, D. L. **Preparation of a gluten-free savoury biscuit with fibres.** Porto Alegre, 2012. 65 f. Monograph (Degree in Food Engineering). Federal University of Rio Grande do Sul, Rio Grande do Sul.

MOURA, C. C. de; PETER, N.; SCHUMACKER, B. de O.; BORGES, L. R.; HELBIG, E. Biscuits enriched with brown linseed meal (Linum usitatissiumun L.): nutritional value and acceptability. **Demetra: Food, Nutrition & Health**, v.9, n. 1, p. 71-81, 2014.

NASCIMENTO, E. M. da G. C. do. **Production of corn products enriched with grains and semi-fatted sesame cake by thermoplastic extrusion.**

Dissertation (master's degree) - Federal Rural University of Rio de Janeiro, Postgraduate Course in Food Science and Technology. 2010.

OLIVEIRA, D. M.; FIRMINO, P. DE T.; MARQUES, D. R., KWIATKOVSKI, A.,GIRIBONI, A. R. M.; SILVA, A. C.; SOUSA, J. DOS S. Physico-chemical characterisation of sesame cv. CNPA-G4 co-products (oil and cake). **Revista Tecnológica**, Special Edition V Simpósio de Engenharia, Ciência e Tecnologia de Alimentos, pp. 37-42, 2011.

OLIVEIRA, M. E.; NABESHIMA, E. H.; CLERICI, M. T. P. S. Sensory evaluation and technological characteristics of cookies developed with defatted sesame flour and resistant starch. **Revista Agrotecnologia,** Anápolis, v. 5, n. 1, p. 115 - 128, 2014.

OLIVEIRA, N. M. A. L.; MACIEL, J. F.; LIMA, A. S.; SALVINO, E. M.; Maciel,C. E. P.; Oliveira, D. P. M. N. Physico-chemical and sensory characteristics of bread enriched with whey protein concentrate and calcium carbonate. **Revista Instituto Adolfo Lutz**. São Paulo, v. 70, n. 1, 2011.

ORDÓÑEZ J. A. **Tecnologia de alimentos: alimentos de origem animal.** Porto Alegre: Artmed; 2005. 280 p.

ORTOLAN, F.; HECKTHEUER, L. H.;MIRANDA, M. Z. de. The effect of storage at low temperature (-4 °C) on the colour and acidity content of wheat flour. **Food Science and Technology,** v. 30, n.1, pp. 55-59, 2010.

PALCZAK, J. et al. Sensory complexity and its influence on hedonic responses: A systematic review of applications in food and beverages, **Food Quality and Preference,** v. 71, p. 66-75, 2019.

PEREZ, P. M. P.; GERMANI, R. Preparation of savoury biscuits with high dietary fibre content using aubergine flour (Solanum melongena, L.). **Ciência Tecnologia de Alimentos**, v. 27, n.1, pp. 186-192, 2007.

PIRES, P. DE S.,QUADROS, G. S. L.,GADELHA, G. G . P. Development and characterisation of gluten-free bread based on vegetable flour. E-xacta, 11(1),

85-95, 2018.

PURLIS, E. Bread baking: technological considerations based on process modelling and simulation. **Journal of Food Engineering,** 103(1), 92-102, 2011.

PYLER, E.J. **Baking science & technology**. 2nd ed. Chicago: Siebel Publ., v.1, p.121-163, 1982.

QUEIROGA, V.P.; SILVA, O. R. R. F. **Technologies used in mechanised sesame cultivation**. Campina Grande: Embrapa Cotton, 2008. 142p. (Embrapa Cotton. Documents, 203).

SANTOS, C. A.; RIBEIRO, R. C.; SILVA, E. V. C.; SILVA, N. S.; SILVA, B. A.;SILVA, G. F.; BARROS, C. V. Elaboration of buriti flour biscuit (Mauritia flexuosa L.f) with and without the addition of oats (Avena sativa L.). **Revista Brasileira de Tecnologia Agroindustrial**, 5: 262-273, 2011.

SANTOS, C . M.,ROCHA, D . A .,MADEIRA, R . A . V.,QUEIROZ, E . DE R.,MENDONÇA, M. M.,PEREIRA, J.,ABREU, C. M. P. Preparation, characterisation and sensory analysis of wholemeal bread enriched with papaya by-product flour. **Brazilian Journal Food Technology,** 21, 1-9, 2018.

SHENOY H., A.; PRAKASH, J. Wheat bran (Triticum aestivum): composition, functionality and incorporation in unleavened bread. **Journal of Food Quality.** v.25, n.3, p. 197 - 211, 2007.

SILVA, C. B. **Effect of the addition of xylanase, glucose oxidase and ascorbic acid on the quality of whole grain wheat flour bread.** 2007. 149f. Dissertation (Master's in Food Technology) - Faculty of Food Engineering, State University of Campinas, Campinas.

SILVA, D. R. S. (2015). **Sesame processing: oil extraction and waste utilisation for food production**. Thesis (PhD) - Postgraduate Programme in Process Engineering, Federal University of Campina Grande, 2015.

SILVA, D. R. S., PESSOA, T.,GURJAO, F. F.,CAVALCANTI MATA, M. E. R. M.,DUARTE, M. E. M. (2017). Use of sesame residue flour in the preparation of savoury biscuits. **Agricultural Technology & Science**, 11, 63-68.

SILVA, D. R. S., PESSOA, T., GURJÃO, F. F., MATA, M. E. R. M. C., & DUARTE,M. E. M. Influence of the incorporation of sesame residue flour on bread quality. **Research, Society and Development,** v. 9, n. 11, p. e46191110108- e46191110108, 2020.

SILVA, F. A. S.; AZEVEDO, C. A. V. de. A new version of the assistat-statistical assistance software. In: World Congress on Computers in Agriculture, 4, Orlando-FL- USA: Proceedings...Orlando: American Society of Agricultural Engineers, p.393-396, 2010.

SILVA, F. A. S; DUARTE, M. E. M; CAVALCANTI MATA, M. E. R. M. Nova methodology for interpreting food sensory analysis data. **Revista Engenharia Agrícola**. v.30, n.5, p.967-973, 2010.

SILVA, I. G. DA, ANDRADE, A. P. C. DE ., SILVA, L. M. R. DA, & GOMES, D. S.
Preparation and sensory analysis of a cookie made from avocado pit flour. **Brazilian Journal of Food Technology**, 22, e2018209, 2019.

SILVA, L. H. DA, PAUCAR-MENACHO, L. M., VICENTE, C. A.,SALLES, A. S.,STEEL, C. J. (2009). Development of flatbreads with the addition of okara flour. **Brazilian Journal of Food Technology**, 12(4), 315-322.

SOUZA, J. M. DE; ÁLVARES, V. DE S.; LEITE, F. M. N.; REIS, F. S.; FELISBERTO, F. A. V. Physico-chemical characterisation of flours from cassava varieties used in the Juruá valley, Acre. **Acta Amazônica**. v. 38, n. 4, p. 761-766, 2008.

STIKIC, R.;GLAMOCLIJA,D.;DEMIN,M.; VUCELIC-RADOVIC, B.; JOVANOVIC, Z.; MILOJKOVIC-OPSENICA, D.; JACOBSEN, S.-E.; MILOVANOVIC, M. Agronomical and nutritional evaluation of quinoa seeds

(Chenopodium quinoa Willd.) as an ingredient in bread formulations. **Journal of Cereal Science**, v. 55, n. 2, p. 132-138, 2012.

STONE, H. S.; SIDEL, J. L. **Sensory evaluation practices**. San Diego: Academic Press, 1993. 308p.

SUBRAMANIAN, N. Technology of vegetable protein foods. **Journal Food Science Technology**, 17, Jan/April 1980.

TAKEITI, C. Y.; CARVALHO, C. W. P.; ASCHERI, J. L. R.; FREITAS, D. G.; technological characteristics of cookies developed with defatted sesame flour and resistant starch. **Revista Agrotecnologia,** Anápolis, v. 5, n. 1, p. 115 - 128, 2014.

WANDERLEY, R. O. S., WANDERLEY, P. A., SILVA, W. A., PAIVA, A. C. C., &OLIVEIRA, J. P. M. Physico-chemical characterisation of French bread enriched with sesame flour Sesamum indicum L. In S. R. Cruz (Ed.), **Anais...** III Simpósio Nacional de Estudos para a Produção Vegetal no Semiárido (pp. 1-4). Campina Grande: Realise Editora, 2018.

ZUNIGA, A.D.G.;COELHO, A.F.S.; FERREIRA, E.S.;RESENDE, E.A.; ALMEIDA, K.N. Shelf life evaluation of wholemeal cashew nut biscuits. **Revista Brasileira de Produtos Agroindustriais,** Campina Grande, v. 13, n.3, p. 251-256, 2011.

yes
I want morebooks!

Buy your books fast and straightforward online - at one of world's fastest growing online book stores! Environmentally sound due to Print-on-Demand technologies.

Buy your books online at
www.morebooks.shop

Kaufen Sie Ihre Bücher schnell und unkompliziert online – auf einer der am schnellsten wachsenden Buchhandelsplattformen weltweit! Dank Print-On-Demand umwelt- und ressourcenschonend produzi ert.

Bücher schneller online kaufen
www.morebooks.shop

MIX
Papier aus verantwortungsvollen Quellen
Paper from responsible sources
FSC® C105338

Printed by Books on Demand GmbH, Norderstedt / Germany